THE LITTLE BOOK OF BOILER FAULT FINDING

Al Craig

Copyright © 2021 Al Craig

All rights reserved

The characters and events portrayed in this book are fictitious. Any similarity to real persons, living or dead, is coincidental and not intended by the author.

No part of this book may be reproduced, or stored in a retrieval system, or transmitted in any form or by any means, electronic, mechanical, photocopying, recording, or otherwise, without express written permission of the publisher.

ISBN: 9798541796612

Cover design by: Art Painter
Library of Congress Control Number: 2018675309
Printed in the United States of America

INTRODUCTION

Hello and welcome to this little book of central heating boiler fault finding.

First of all a little about myself. I've worked in the heating industry for nearly 20 years; servicing, repairing, and installing boilers. I've also worked on a technical support desk, as well as being involved with training of apprentices. So it's safe to say I've a fair bit of experience and I've also learned how to help you to help yourself.

By simply telling somebody the answers or just guessing, you never actually learn or improve your knowledge. So I've tried to give you the foundations of what to look for and how to fault find for yourselves.

WHO IS THIS BOOK FOR?

This booked is primarily aimed at people working in the heating industry, both qualified engineers and apprentices. So I've assumed you have some gas knowledge and won't be covering all of the basics here.

It's not aimed at DIY'er's and as always I'd advise you to get any gas work done by a qualified Gas Safe registered engineer.

Hopefully you'll learn a few tricks of the trade as the book progresses, but after reading hang onto a copy in your van, you never know when it may come in handy.

My aim for writing this book, is to help you get it right, at the lowest possible cost for you and your customers.

FIRST STAGE FAULTFINDING

One of the biggest mistakes people make when searching for faults is going too deep, too quickly.

Good faultfinding requires paying attention to all the free information available. This includes:

- Interrogating the customer
- Looking (are there fault lights, demand indicators)
- Listening (Can you hear any noise, such as fan or pump)
- Checking all demands are calling.

Frequently diagnosis can be reached without even taking the case off an appliance.

BASIC CHECKS

After carrying out a thorough first stage fault finding, it's time to look at basic checks.

Again this sometimes saves us lots of time and potential money on unneeded parts.

Theres a full guide on how to carry out basic checks, later in this book, but these include.

- Inlet pressure/ working pressure, gas rate
- Blip test
- Water pressure, system bled of air
- Electrical checks including polarity
- System checks - pressure, controls, flow and return temperatures, all system valves open
- Flue checks
- Case seals

- Condensate trap and discharge pipework clear

ELECTRICAL CHECKS

It's important when we start any faultfinding that we've checked all our basics.

I'd frequently spend 30 minutes on the phone with somebody searching for a fault, where lots of parts had already been fit, and it turned out to be something simple, that could've been rectified for nothing.

It's a good idea to do the following checks before you get too involved. You can do this as part of your safe isolation, to ensure you stay safe whilst you've got your hands in a boiler.

Short circuit test

This is simply your main and switched lives to neutral, we're looking for either a readable resistance, but on newer boilers it's more common to get a high reading, this is fine. We just need to make sure we're not getting continuity, as this indicates a short circuit.

Earth continuity

This is simply checking you have an earth path back to the fuse spur, using your multimeter on ohms check the main earth to all visible earths and back to the fuse spur, it's good practice to check it back to the screw on the spur.

A faulty earth can be one of the main causes of rectification problems.

Resistance to earth

Here we're checking you main live to earth, we're look for open circuit, any readable resistances should be investigated.

Polarity

This is another which catches people out frequently, I've seen people fit £300 pcbs, thinking as it was all dead it had to be that, when it turned out it was just a broken Neutral.

Check L-N 240v, L-E 240v, N-E Up to 15v

It's a good habit to get into to check all these before you dive in at the deep end. It'll probably save you loads of time and hopefully money on wrong diagnosis.

SECOND STAGE FAULT FINDING

Sequence of operation

All boilers work slightly differently, however most have a similar basic principle.

Frequently you'll find a sequence of operation in the M.Is. If you can, this saves you time as it can identify where it's falling down.

However if there's no sequence available, it's useful to look to the wiring diagram, or components you can see, and try and put a sequence together in your head.

Most modern boilers have a similar principle in how they work, once you know a basic sequence you can tailor it to your needs.

Here is generic sequence that will help aid our second stage faultfinding.

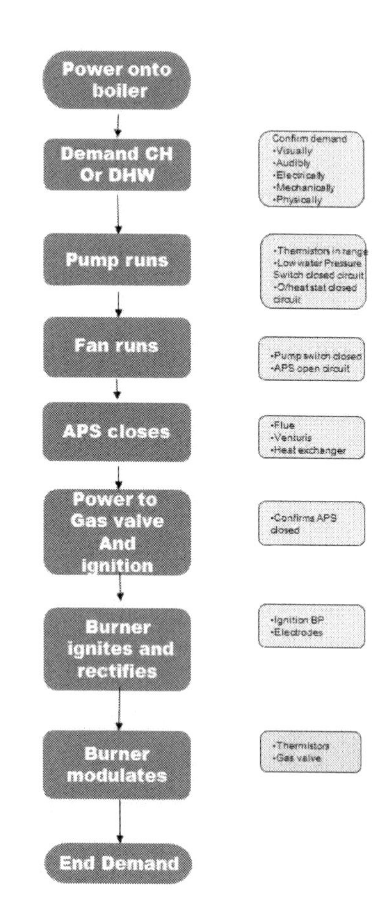
Power onto boiler
Demand CH Or DHW
Confirm demand
•Visually
•Audibly
•Electrically
•Mechanically
•Physically
Pump runs
•Thermistors in range
•Low water Pressure Switch closed circuit
•O/heat stat closed circuit
Fan runs
•Pump switch closed
•APS open circuit
APS closes
•Flue
•Venturis
•Heat exchanger
Power to Gas valve And ignition
•Confirms APS closed
Burner ignites and rectifies
•Ignition BP
•Electrodes
Burner modulates
•Thermistors
•Gas valve
End Demand

WIRING DIAGRAMS

It's worth trying to locate a wiring diagram for the boiler, these are usually in the manufacturer's instructions, if you can't find it, try searching online.

The wiring diagram is one of the most useful pieces of information you can get hold of, we don't have to be electronics experts to use them, but it does give us an indication as to what components the boiler has, and using our knowledge of basic sequences of operation can help us to identify where our problem is.

For detailed information on how to read and understand wiring diagrams, consult an electronics book, as this is outside the scope of this book.

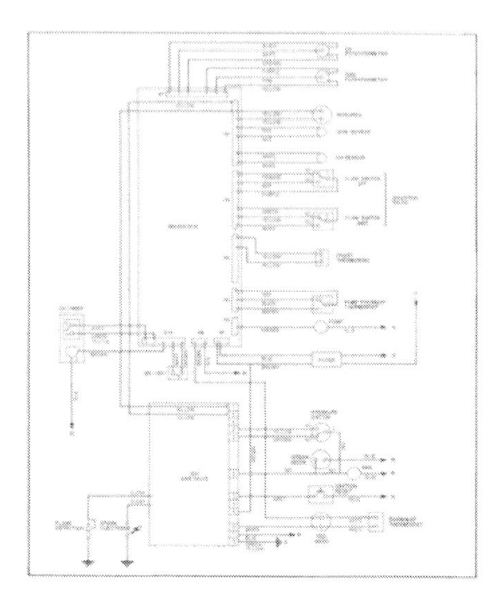

An example of a wiring diagram, showing us what components the boiler has and how they work in conjunction with each other.

GAS CHECKS

There are a few gas related checks which are useful at telling us what's going on with our boiler.

Inlet pressure check

Putting a gauge on the inlet test point will tell us the standing pressure at the appliance, when we put it into operation, this should drop to 21+-2, if it's outside these tolerances, this should be checked at the meter to identify if it's a fault on the meter regulator or if it's related to gas pipe size.

Another useful check whilst the gauge is connected is a **blip test**. This is used when the boiler starts a sequence but doesn't light. Check the standing pressure and look for a drop when the boiler tries to fire, if it doesn't move, the gas valve isn't opening, this can be used to aid faultfinding, if there's voltage at the gas valve, but its not opening, then we have a faulty gas valve.

Burner pressure

Older appliances have a set burner pressure, usually a high and low if there's a modureg fit, such as on a combi. It's important to check all gas pressures against M.Is as an incorrectly set gas valve can create lock out issues or poor/ unsafe combustion.

Gas rate

On pre mixed burners, we need to take a gas rate, this will tell us how much gas the boiler is using, make sure you follow the M.Is as some have special service modes than need following.

This can be especially useful at proving if there is actually a fault, I've visited numerous 24KW boilers in the middle of winter struggling to heat a bath of water, but could prove there was nothing wrong with the boiler, it was just a simple issue of the flow rate being too high.

PRE MIX BURNERS

Since 2005 all boilers fitted in the UK had to be of a condensing type, this meant most boilers moved to a forced pre mix burner from a naturally aspirated type.

The obvious benefit is an improvement in efficiency, but whilst they are clearly more efficient, there are few important things to remember.

Always check the performance tester readings and make sure you compare them to the manufactures instructions, these may include specific High and Low CO/CO2 readings. I cant stress enough how important it is to check these, so often I've spoken to people who checked the readings and adjusted the gas valve as they where out. **Do not do this.**

By adjusting the gas valve you are masking the issue, the readings are probably out due to lack of servicing, if your readings are out the boiler should be stripped and cleaned before touching the gas valve, but as always follow the M.Is where in doubt.

Another test that needs doing on a pre mix appliance is gas rate. As there is no way of checking a burner pressure, it's important to check the appliance is burning the correct amount of gas. This should be checked against the data badge or M.Is, anything that is more than 10% over or 5% less than it should be, needs investigating as it could identify a safety or performance issue.

For example a 30 KW combi gas rating at 20KW wont get the water as hot as it should, this may be caused by a restricted main heat exchanger as the restriction will act as a damper to the fan, reducing the amount of gas its able to pull through the heat exchanger.

Likewise a 30 KW boiler gas rating at 40KW may indicate an incorrectly set gas valve or parameter on the pcb.

Some appliances have specific fan pressure readings, this will indicate if a strip and clean is required.

Where in doubt always contact the manufacturers technical support for guidance.

COMPONENT RESISTANCES

Most components have what we call a generic reading, this means when we're faultfinding, we can look for an approximate resistance range to give us an idea as to whether the component is OK.

To carry out a resistance check, ensure the power is off, trace the wires from the component back to the pcb, unplug the connector and check the resistance using the ohms scale on your multimeter.

COMPONENT READINGS

Fans (standard fixed speed)

Expect 25 – 90 ohms, with no power the fan should spifreely when pushed.

Most pre mixed boilers have fans that include a PCB, these usually can't be checked for resistance. In this case it's best to consult manufacturers instructions.

Pumps

Generally between 150 – 350 ohms. Check for water leaks internally, remove blead point does it spin freely.

Gas Valve

Different types can have different readings, ranging from 80 ohms to 5 KOhms. We're looking for a readable reading, so a short circuit (continuity) or open circuit (O/L) would indicate an issue.

If a gas valve has 5 pins, the earth is generally the middle and the 2 solenoids are either side, this is where we'd be looking for a readable resistance.

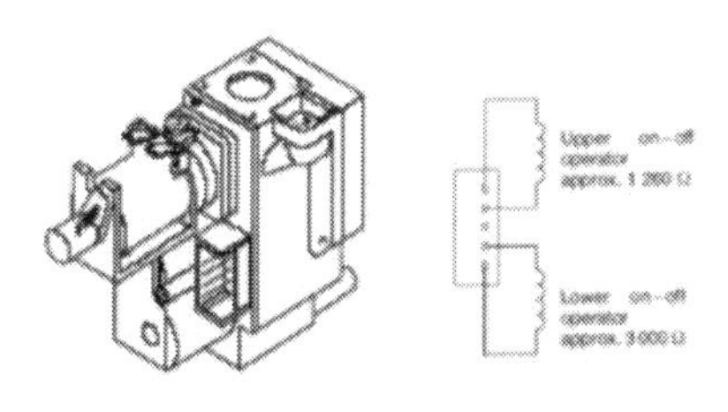

Sensors

There's a full section on these, but generally 1.5 KOhms to 15 Ohms.

Overheat stats/ other safety stats

These are generally open or closed circuit and we shouldn't normally get a readable resistance, an open circuit would indicate something may have tripped.

Air pressure switches

these are usually 2 or 3 wires, for 2 wire stats it should read OL at rest and continuity when made, take great care when making an APS manually as blowing down them can damage the delicate diaphragm inside.

For 3 wire switches we're looking for continuity between C-NC at rest and continuity between C-NO when made, it's worth testing all 3 wires in case the switch has failed internally, you should get OL between C and the opposite wire.

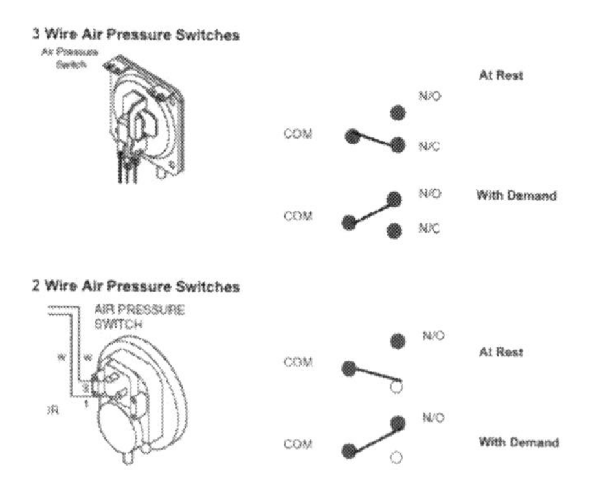

Flow switches

Generally 2 wire, at rest should have O/L between the connections. When made should have continuity, best checked from pcb to prove wires.

Flow sensors (hall sensors)

These are found on most modern combi boilers, easily identified by having 3 wires.

To test these hall sensors, first check for voltage DC at each wire at the pcb with no water running. You should get 1 input with a set voltage on, 1 ground wire with 0 on and another with either 0 or the same as the input voltage on. When we run a tap, repeat this test and you should now have a change in voltage on 1 wire, this should be around half the voltage of the input wire, if it's not then we most likely have an issue with our flow switch.

PROVING FAILED COMPONENTS

It's possible to prove many components by simply bridging the components out, you need to use your own judgement as to whether you can do this safely.

I used to use some clip on fly leads as a means on linking overheat stats and air pressure switches out, you could also use electricians pliers, again make sure these are suitable prior to using them live.

TEMPERATURE TESTS

Frequently we'll have issues with circulation in systems, it's important to have a set of temperature clamps to help prove where a fault may lie, usually these can plug into your performance tester or multimeter.

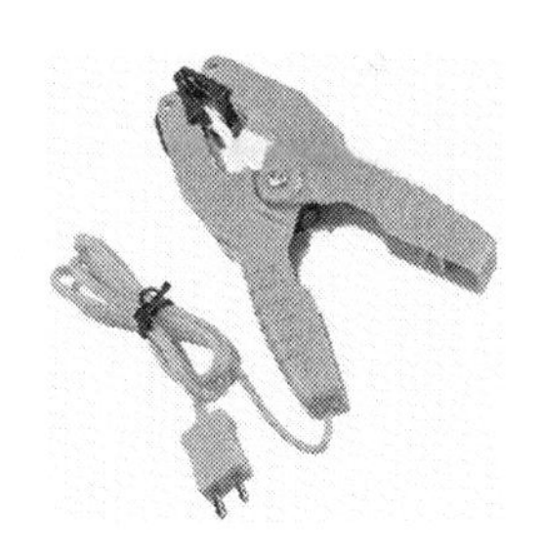

THERMISTOR READINGS

It's helpful to know what a boiler's thermistor readings are, as this is what's telling the pcb what it needs to do.

These generally range from 15 Kohm to 1.5 Kohm. It's important to test these against the temperature they are currently reading. For example if a sensor is reading 15Kohm and the boiler is at room temperature, we can assume its about right, however if the boiler is at 82 degrees C and you're still getting 15KOhm, then you may have an issue.

To test thermistors, we should trace the wires back to the pcb, unplug the connector and measure the resistance using the Ohms scale on our multimeter. We can than confirm this by attaching a temperature clamp as near to the sensor as we can. **NB. This is a dead test**

It's good practice to test back at the pcb, as this will test the integrity of the leads also, and give a true reading as to what the pcb is actually reading.

Once we have our reading we can compare it to our chart, see figure X and see if this is where our problem lies.

If the appliance you're working on has more than one sensor, ie some combis may have 3, its a good idea to check all of them as they should all be in a similar range.

Any strange readings or readings that are jumping around would usually indicate an issue with one of the sensors. As always, if you notice any odd readings, its a good idea to touch your multimeter leads together and check you're only getting continuity, around 5 ohms.

Top tip...

It's worth saving a few old thermistors from old boilers etc and keeping them in your toolbox, you can often plug these in dry whilst faultfinding to prove a faulty sensor.

NTC Sensor Resistances

Temperature in degrees C	Resistance in KOhms
0	32
5	25
10	19
15	15
20	12
25	10
30	8
35	6
40	5
45	4
50	3.5
55	3
60	2.5
65	2
70	1.7
75	1.5
80	1.2
85	1
90	0.9
95	0.7
100	0.6

PROVING HOT WATER PERFORMANCE

Again a frequent issue, this usually rears its head in the winter when the hot water is being run too fast for the boiler to cope.

Simply start by checking the temperature of a cold tap, then check the manufacturers instructions, it should give a temperature rise for a desired flow rate.

You'll need a weir gauge here, measure the water flow rate and if necessary adjust it to the specified flow rate, then check the water temperature rise, you'll frequently find this a problem on older lower output boilers.

Once your hot water flow rate is set per the M.I's, it'll give an indication as to where, if applicable, you need to look next.

If the boiler gas rate is correct but the hot water performance is still inadequate, we need to check the central heating flow pipe.

If you temperature clamp highlights a rise in temperature, despite the heating not been on, then perhaps we have a faulty diverter valve.

If this isn't the case, the next steps are to look for an internal bypass, which may have failed, causing the boiler to circulate round the boiler only.

There is also the potential the plate heat exchanger is restricted, a tell tale sign of this is the water going hot, cold, hot cold etc...we look at how you can prove this later on using your temperature clamps.

TESTING FOR FLOW ACROSS A MAIN HEAT EXCHANGER

Place your clamps inside the boiler case on the flow and return to the main heat exchanger and you should be looking for a differential of around 10-20 degrees, depending on the boiler.

If it seems ok on heating but you have an issue with your hot water, simply run the hot water tap, on a combi, this will measure the resistance over the plate heat exchanger, any more than 20 degrees highlights a restriction, you'll frequently find the hot water going hot and cold if this is a problem.

SYSTEMS

System faults are outside the scope of this book, however, it remains to be said, when encountering any boiler issue it's important to look at the whole of the system for ideas as to what else could be going on.

For example, if you replace a plate heat exchanger due to it having a high temperature differential, caused by a blockage, you need to look at the system for the whole picture.

If the water is black and full of sludge, it's important this is cleared before fitting parts, as you'll get the same issue recurring time and time again.

GOOD SYSTEM GUIDELINES

I always like to inspect the full system before I start replacing parts, especially on system boilers, before we even touch a system boiler we need to ascertain the basics:

- Are all radiators and system bled?
- Is there pressure in the system/ or F&E tank full on an open system?
- Can you hear the pump?
- Have the motorised valves powered over?
- Are you getting switch live (on system boilers)

’S ’Plans and ‘Y ’plans

Theres numerous configurations of systems, but I’m only going to include the main two, these are Y plans and S plans, they both work slightly different, and as I said I'm not going into depth, as this book is about faultfinding on boilers, however it helps to have a good understanding to ensure we can confirm if a boiler is getting a switched live demand.

S PLANS

Grey = Permanent live
Blue = Neutral
Y/G = Earth
Orange = Pump/boiler live
Brown = switch live from room stat or cylinder stat

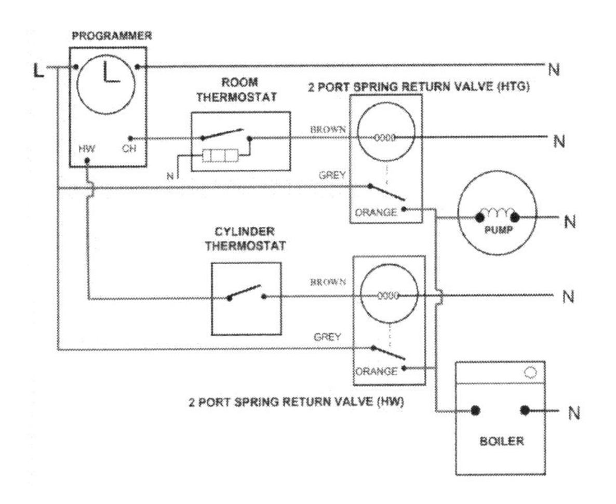

Y PLANS

Blue = Neutral
Y/G = Earth
Orange = Pump/ boiler live
White or Brown = Central heating on
Grey = Hot water off

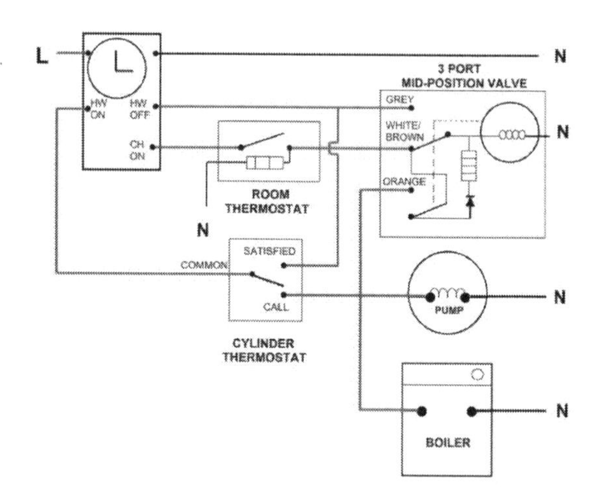

USEFUL HINTS AND TIPS

- Turn off any pre heats to ensure we are testing the operation of any flow switches

- Have a set of pre made jumper leads in your tool kit, these can be invaluable for bridging components out, such as timers or safety stats to prove a fault.

- 3 wire APS on some older boilers can be bridged by linking all 3 wires together.

- If you suspect an airflow issue try removing the combustion case, this will usually generate a little more air flow and help to make the APS.

- Most of the time faults can be resolved by simply stripping down and servicing an appliance, choked heat exchangers, restricted fans, bent electrodes are all easy fixes that are often overlooked.

- Expansion vessels can frequently be repaired, simply by draining the condensation from them. Simply drain the boiler, remove the Schrader core, put some tubing over the valve, refill the boiler to 1.5 bar and the water pressure should drain any condensation stuck inside the air chamber, it also proves if the vessel has blown internally. Refit the Schrader core and refill the pressure to 1 bar, or per the M.I.

USEFUL LINKS

Gas safe

Worcester Bosch

Ideal

Baxi

Vailaint

Glow worm

Viessmann

Vokera

Biasi

...AND FINALLY

Thanks for reading this little book on faultfinding.

I've purposely kept this book brief, instead of adding lots of filler that's not relevant to your day to day work.

I'll be updating this with new ideas and information as it comes.

Thanks again for reading, and remember stay safe!